Notebook no. _____

Issued to: _____

Date issued: _____

School: _____

Class: _____

Returned on: _____

© 2018 Bio-Rad Laboratories, Inc. All rights reserved. 10000104110 (1661051) Ver A

Guidelines and Instructions

1. Your laboratory notebook is a permanent record of your laboratory work.

2. Your entries should be made in permanent ink.

3. Record the project title when starting each new page.

4. Include all data, notes, diagrams, ideas, calculations, photographs, plans, observations, etc. used in your experiments. Include enough details so that a colleague can continue your work from the information provided.

5. All corrections should be done by drawing a single line through the incorrect data, with dates and initials.

6. Date and sign all entries you make. A witness or witnesses who understand your work should also sign and date your entries.

Table of Contents	Page No.

Table of Contents	Page No.

Table of Contents	Page No.

Table of Contents | Page No.

Table of Contents	Page No.

Table of Contents

	Page No.

Table of Contents	Page No.

Table of Contents	Page No.

1

Title: Project no.
 Book no.

From page no.

To page no.

Witnessed and understood by me: | Date: | Invented/recorded by: | Date:

2

Title:

Project no.

Book no.

From page no.

To page no.

Witnessed and understood by me: | Date: | Invented/recorded by: | Date:

Title: Project no.
         Book no.

From page no.

To page no.

Witnessed and understood by me: Date: Invented/recorded by: Date:

4

Title:

Project no.

Book no.

From page no.

To page no.

Witnessed and understood by me: Date: Invented/recorded by: Date:

6

Title:

Project no.

Book no.

From page no.

To page no.

Witnessed and understood by me: | Date: | Invented/recorded by: | Date:

8

Title:

Project no.

Book no.

From page no.

To page no.

| Witnessed and understood by me: | Date: | Invented/recorded by: | Date: |

9

Title: Project no.
        Book no.

From page no.

To page no.

Witnessed and understood by me: | Date: | Invented/recorded by: | Date:

10

Title:

Project no.

Book no.

From page no.

To page no.

Witnessed and understood by me: | Date: | Invented/recorded by: | Date:

11

12

Title:

Project no.

Book no.

From page no.

To page no.

Witnessed and understood by me: | Date: | Invented/recorded by: | Date:

14

Title:

Project no.

Book no.

From page no.

To page no.

Witnessed and understood by me: | Date: | Invented/recorded by: | Date:

16

Title:

Project no.

Book no.

From page no.

To page no.

Witnessed and understood by me: | Date: | Invented/recorded by: | Date:

17

18

Title:

Project no.

Book no.

From page no.

To page no.

Witnessed and understood by me: | Date: | Invented/recorded by: | Date:

20

Title:

Project no.

Book no.

From page no.

To page no.

Witnessed and understood by me: | Date: | Invented/recorded by: | Date:

21

Title:

Project no.

Book no.

From page no.

To page no.

Witnessed and understood by me: | Date: | Invented/recorded by: | Date:

Title:

Project no.

Book no.

From page no.

To page no.

Witnessed and understood by me: Date: Invented/recorded by: Date:

23

Title:

Project no.

Book no.

From page no.

To page no.

Witnessed and understood by me: | Date: | Invented/recorded by: | Date:

24

Title:

Project no.

Book no.

From page no.

To page no.

Witnessed and understood by me: | Date: | Invented/recorded by: | Date:

26

Title:

Project no.

Book no.

From page no.

To page no.

Witnessed and understood by me: | Date: | Invented/recorded by: | Date:

27

28

Title:

Project no.

Book no.

From page no.

To page no.

Witnessed and understood by me: | Date: | Invented/recorded by: | Date:

30

Title:

Project no.

Book no.

From page no.

To page no.

Witnessed and understood by me: Date: Invented/recorded by: Date:

32

Title:

Project no.

Book no.

From page no.

To page no.

Witnessed and understood by me: | Date: | Invented/recorded by: | Date:

33

34

Title:

Project no.

Book no.

From page no.

To page no.

Witnessed and understood by me: | Date: | Invented/recorded by: | Date:

35

36

Title:

Project no.

Book no.

From page no.

To page no.

Witnessed and understood by me: | Date: | Invented/recorded by: | Date:

37

39

Title:

Project no.

Book no.

From page no.

To page no.

| Witnessed and understood by me: | Date: | Invented/recorded by: | Date: |

40

Title:

Project no.

Book no.

From page no.

To page no.

Witnessed and understood by me: | Date: | Invented/recorded by: | Date:

41

42

Title:

Project no.

Book no.

From page no.

To page no.

Witnessed and understood by me: | Date: | Invented/recorded by: | Date:

44

Title:

Project no.

Book no.

From page no.

To page no.

Witnessed and understood by me: Date: Invented/recorded by: Date:

46

Title:

Project no.

Book no.

From page no.

To page no.

Witnessed and understood by me: | Date: | Invented/recorded by: | Date:

47

48

Title:

Project no.

Book no.

From page no.

To page no.

Witnessed and understood by me: | Date: | Invented/recorded by: | Date:

50

Title:

Project no.

Book no.

From page no.

To page no.

Witnessed and understood by me:	Date:	Invented/recorded by:	Date:

52

Title:

Project no.

Book no.

From page no.

To page no.

Witnessed and understood by me:　　Date:　　Invented/recorded by:　　Date:

53

Project no.

Book no.

From page no.

To page no.

Witnessed and understood by me: | Date: | Invented/recorded by: | Date:

54

Title:

Project no.

Book no.

From page no.

To page no.

Witnessed and understood by me: | Date: | Invented/recorded by: | Date:

56
Title:

Project no.
Book no.

From page no.

To page no.

Witnessed and understood by me: | Date: | Invented/recorded by: | Date:

57

58

59

60

Title:

Project no.

Book no.

From page no.

To page no.

Witnessed and understood by me: | Date: | Invented/recorded by: | Date:

62

Title:

Project no.

Book no.

From page no.

To page no.

| Witnessed and understood by me: | Date: | Invented/recorded by: | Date: |

64

65

68

Title:

Project no.

Book no.

From page no.

To page no.

Witnessed and understood by me: Date: Invented/recorded by: Date:

70

Title:

Project no.

Book no.

From page no.

To page no.

Witnessed and understood by me: | Date: | Invented/recorded by: | Date:

71

72

73

Title: Project no.
Book no.

From page no.

To page no.

Witnessed and understood by me: Date: Invented/recorded by: Date:

74

76

Title: Project no.
Book no.

From page no.

To page no.

Witnessed and understood by me: | Date: | Invented/recorded by: | Date:

78

79

Project no.
Book no.

From page no.

To page no.

Witnessed and understood by me: | Date: | Invented/recorded by: | Date:

80

Title:

Project no.

Book no.

From page no.

To page no.

Witnessed and understood by me: Date: Invented/recorded by: Date:

82

Title:

Project no.

Book no.

From page no.

To page no.

Witnessed and understood by me: | Date: | Invented/recorded by: | Date:

85

Project no.

Book no.

From page no.

To page no.

Witnessed and understood by me: | Date: | Invented/recorded by: | Date:

86

Title:

Project no.

Book no.

From page no.

To page no.

Witnessed and understood by me: | Date: | Invented/recorded by: | Date:

87

Title:

Project no.

Book no.

From page no.

To page no.

| Witnessed and understood by me: | Date: | Invented/recorded by: | Date: |

88

Title:

Project no.

Book no.

From page no.

To page no.

Witnessed and understood by me: | Date: | Invented/recorded by: | Date:

89

Project no.

Book no.

From page no.

To page no.

Witnessed and understood by me: | Date: | Invented/recorded by: | Date:

90

Title:

Project no.

Book no.

From page no.

To page no.

Witnessed and understood by me: | Date: | Invented/recorded by: | Date:

91

93

Title:

Project no.

Book no.

From page no.

To page no.

Witnessed and understood by me: | Date: | Invented/recorded by: | Date:

94

Title:

Project no.

Book no.

From page no.

To page no.

Witnessed and understood by me: | Date: | Invented/recorded by: | Date:

96

Title:

Project no.

Book no.

From page no.

To page no.

Witnessed and understood by me: Date: Invented/recorded by: Date:

100

Title:

Project no.

Book no.

From page no.

To page no.

Witnessed and understood by me: | Date: | Invented/recorded by: | Date:

101

102

104

Title:

Project no.

Book no.

From page no.

To page no.

Witnessed and understood by me: | Date: | Invented/recorded by: | Date:

105

106

Title:

Project no.

Book no.

From page no.

To page no.

Witnessed and understood by me: | Date: | Invented/recorded by: | Date:

108

Title:

Project no.

Book no.

From page no.

To page no.

Witnessed and understood by me: | Date: | Invented/recorded by: | Date:

110

111

Project no.

Book no.

From page no.

To page no.

Witnessed and understood by me: | Date: | Invented/recorded by: | Date:

112

Title:

Project no.

Book no.

From page no.

To page no.

Witnessed and understood by me: | Date: | Invented/recorded by: | Date:

113

114

Title:

Project no.

Book no.

From page no.

To page no.

Witnessed and understood by me: | Date: | Invented/recorded by: | Date:

116
Title:

Project no.

Book no.

From page no.

To page no.

Witnessed and understood by me: | Date: | Invented/recorded by: | Date:

117

118

Title:

Project no.

Book no.

From page no.

To page no.

Witnessed and understood by me: | Date: | Invented/recorded by: | Date:

120

Title:

Project no.

Book no.

From page no.

To page no.

Witnessed and understood by me: | Date: | Invented/recorded by: | Date:

126

Title:

Project no.

Book no.

From page no.

To page no.

Witnessed and understood by me: | Date: | Invented/recorded by: | Date:

134

Title:

Project no.

Book no.

From page no.

To page no.

Witnessed and understood by me: | Date: | Invented/recorded by: | Date:

137

138

139

140

Title:

Project no.

Book no.

From page no.

To page no.

Witnessed and understood by me: | Date: | Invented/recorded by: | Date:

142

143

144

Title:

Project no.

Book no.

From page no.

To page no.

Witnessed and understood by me: | Date: | Invented/recorded by: | Date:

145

146

Title:

Project no.

Book no.

From page no.

To page no.

Witnessed and understood by me: | Date: | Invented/recorded by: | Date:

147

148

Title:

Project no.

Book no.

From page no.

To page no.

Witnessed and understood by me:　　Date:　　Invented/recorded by:　　Date:

149

Title:

Project no.

Book no.

From page no.

To page no.

Witnessed and understood by me: | Date: | Invented/recorded by: | Date:

150

151

152

Title:

Project no.

Book no.

From page no.

To page no.

Witnessed and understood by me: | Date: | Invented/recorded by: | Date:

158

Title:

Project no.

Book no.

From page no.

To page no.

Witnessed and understood by me: | Date: | Invented/recorded by: | Date:

160

Title:

Project no.

Book no.

From page no.

To page no.

Witnessed and understood by me: | Date: | Invented/recorded by: | Date:

162

Title:

Project no.

Book no.

From page no.

To page no.

Witnessed and understood by me: | Date: | Invented/recorded by: | Date:

164

Title:

Project no.

Book no.

From page no.

To page no.

Witnessed and understood by me: | Date: | Invented/recorded by: | Date:

165

Title:
Project no.
Book no.

From page no.

To page no.
165

Witnessed and understood by me: | Date: | Invented/recorded by: | Date:

168

Title:

Project no.

Book no.

From page no.

To page no.

Witnessed and understood by me: | Date: | Invented/recorded by: | Date:

169

170

172

Title:

Project no.

Book no.

From page no.

To page no.

Witnessed and understood by me: | Date: | Invented/recorded by: | Date:

173

174

Title:

Project no.

Book no.

From page no.

To page no.

Witnessed and understood by me: | Date: | Invented/recorded by: | Date:

176

177

178

Title:

Project no.

Book no.

From page no.

To page no.

Witnessed and understood by me: Date: Invented/recorded by: Date:

179

180

Title:

Project no.

Book no.

From page no.

To page no.

Witnessed and understood by me: | Date: | Invented/recorded by: | Date:

181

Title:

Project no.

Book no.

From page no.

To page no.

Witnessed and understood by me: | Date: | Invented/recorded by: | Date:

182

Title:

Project no.

Book no.

From page no.

To page no.

Witnessed and understood by me: | Date: | Invented/recorded by: | Date:

183

Title:

Project no.

Book no.

From page no.

To page no.

Witnessed and understood by me: | Date: | Invented/recorded by: | Date:

184

Title:

Project no.

Book no.

From page no.

To page no.

Witnessed and understood by me: | Date: | Invented/recorded by: | Date:

185

Title:

Project no.

Book no.

From page no.

To page no.

Witnessed and understood by me: | Date: | Invented/recorded by: | Date:

186

Title:

Project no.

Book no.

From page no.

To page no.

Witnessed and understood by me: | Date: | Invented/recorded by: | Date:

187

188

189

190